LABORATORY EXPERIENCES
IN
GROUP THEORY

A manual to be used with
EXPLORING SMALL GROUPS

©1996 by

The Mathematical Association of America, Inc.

Library of Congress Number 96-77787

ISBN 0-88385-705-7

Printed in the United States of America

Current Printing (last digit):
10 9 8 7 6 5 4 3 2 1

LABORATORY EXPERIENCES IN GROUP THEORY

A manual to be used with
EXPLORING SMALL GROUPS

Ellen Maycock Parker
DePauw University

Published by
The Mathematical Association of America

CLASSROOM RESOURCE MATERIALS

This series provides supplementary material for students and their teachers—laboratory exercises, projects, historical information, textbooks with unusual approaches for presenting mathematical ideas, career information, and much more.

Proofs Without Words, by Roger Nelsen
A Radical Approach to Real Analysis, by David Bressoud
She Does Math! Real-Life Problems from Women on the Job, edited by Marla Parker
Learn from the Masters! edited by Frank Swetz, John Fauvel, Otto Bekken, Bengt Johansson, and Victor Katz
101 Careers in Mathematics, edited by Andrew Sterrett

These volumes can be ordered from:
MAA Service Center
P.O. Box 91112
Washington, DC 20090-1112
1-800-331-1MAA   FAX: 1-301-206-9789

Introduction

I began using *Exploring Small Groups* for demonstration purposes in the spring of 1990. One day, instead of answering a question asked by a student in the class, I sent the students to the computer lab to generate some conjectures. The following day, they were asked to prove or disprove several of the conjectures made by their classmates. One student commented, "How can we work on these when we don't know whether they are true?" At that point, I realized that my students had had very little opportunity to experience mathematical discovery.

This laboratory manual is an attempt to remedy this situation. The study of groups is really the study of symmetry. Many of the most important and basic concepts of group theory are demonstrated with geometry and seen through examples. The structure of groups can often be recognized by students in an introductory course if enough examples are available. The computer allows you to investigate examples without the tedium of hand computations. Often a pattern emerges after looking at four or six examples.

Your instructor may have scheduled a special laboratory time for your class or you may be working through these labs as homework assignments. Either approach can be very useful. If your instructor allows it, try to work with another student—one of you can type and the other record the data. You should take care in writing up the assignments. Sometimes you learn more as you go back over the material after the lab. You'll see that almost every lab begins with a problem to be done by hand. I've discovered that this is an important step because the data from the computer can seem pretty mysterious. So, even if it seems tedious, be sure to do this paper and pencil work. Some of the labs don't use the computer at all and the laboratory time is really just a chance to work collaboratively.

A key part of these labs is that you are asked to make conjectures. Many of the labs are more open-ended than the ones typically used in calculus or linear algebra. Use your imagination–you may even come up with a new theorem (or at least something your instructor doesn't know!). This is the way you see what real mathematical research is all about: the uncertainty, the dead ends, and the ultimate satisfying "aha!"

Most students find the laboratory component of this course to be an enjoyable alternative to the "theorem-proof-example" format of a standard course. I hope that you agree at the end of the semester with a student of mine, who wrote on his evaluation of the course:

> I myself am truly grateful for the laboratory component. ... Work on the computer helped to make the abstract theory more concrete. ... One of the best things about the labs was that we formed our own conjectures about the patterns we saw. ... I believe that the progression of (1) lab, (2) conjecture, (3) class discussion, (4) proof was highly beneficial in gaining understanding of the abstract material of the course.

Acknowledgments

This laboratory manual would not have been possible without the support and encouragement of many. I wish to thank Don Albers, Andy Sterrett, Elaine Pedreira, and Beverly Ruedi of the MAA, for recognizing the potential of the project and working through the many concerns and details. I also wish to thank my colleague at DePauw, Woody Dudley, for his careful reading and editing of the original manuscript, Ladnor Geissinger of the University of North Carolina for his help with the software, and Ed Keppelmann of the University of Nevada at Reno for his feedback after using the labs in his classroom. Finally, special thanks are given to the DePauw University students who have enrolled in Mathematics 371 over the past few years and have willingly become the guinea pigs for this interesting project.

Contents

Chapter 1

Groups and Geometry

Group theory is the mathematics of symmetry. You will look at three very basic shapes, the triangle, the rectangle and the square, to see how their symmetries give rise to some interesting mathematics. Then, you'll see how *Exploring Small Groups (ESG)* presents this information. This lab is, therefore, an introduction to group theory and to *ESG*.

1.1   Before the Lab

Definition 1.1: Let S be a set. A **permutation** of S is a function from S to S that is one-to-one and onto.

We'll be using the sets $\{1, 2, 3\}$ and $\{1, 2, 3, 4\}$ for this lab. First, trace and cut out one copy of each shape found in section 4, at the end of this lab. Label each vertex on both sides with the same number. Be sure that your copy duplicates the printed shape exactly, with vertex numbers matching.

Let's start with the triangle. Place your cut-out triangle form exactly on top of the outline of the printed triangle, matching the numbers. Now rotate the triangle 120° counterclockwise. Notice that the vertex labelled "1" on the cut-out triangle is moved to "2" on the fixed triangle. Likewise, "2" on the cut-out triangle is moved to "3" and "3" is moved to "1." We can record this with the notation

$$1 \to 2$$
$$2 \to 3$$
$$3 \to 1.$$

1. Record on your data sheet all the permutations obtained by counterclockwise rotations of the triangle, using the arrow notation as above.

We can also create permutations by "flipping" the triangle through a line of reflection. Think of a line that goes through the vertex labeled with "1" and the midpoint of the side opposite the vertex "1" on the stationary triangle. Now flip the cut-out triangle over this line. This action creates the permutation

$$1 \to 1$$
$$2 \to 3$$
$$3 \to 2.$$

2. Determine all possible flips of the triangle. Record the permutations you created on your data sheet.

3. Did you generate all possible permutations of $\{1, 2, 3\}$ by using the triangle? Why or why not? If you answered no, give an example of a permutation not generated geometrically.

4. Now record on your data sheet the permutations of $\{1, 2, 3, 4\}$ generated by rotations and flips of the rectangle.

5. Record on your data sheet the permutations of $\{1, 2, 3, 4\}$ generated by rotations and flips of the square.

6. Compare the permutations in problems 4 and 5. Are there some that are the same? Explain this geometrically. Did you obtain all possible permutations of $\{1, 2, 3, 4\}$ with these geometric manipulations? Why or why not? If you answered no, give an example of a permutation not generated geometrically.

We will say that one of the rotations or flips of the geometric figure is a **symmetry** of that figure.

Now compare your answers to those given below. For convenience, we should all use the same notation for the permutations.

Symmetries of a triangle

1_T	r_1	r_2
$1 \to 1$	$1 \to 2$	$1 \to 3$
$2 \to 2$	$2 \to 3$	$2 \to 1$
$3 \to 3$	$3 \to 1$	$3 \to 2$

m_1	m_2	m_3
$1 \to 1$	$1 \to 3$	$1 \to 2$
$2 \to 3$	$2 \to 2$	$2 \to 1$
$3 \to 2$	$3 \to 1$	$3 \to 3$

Symmetries of a rectangle

1_R	r_1	m_1	m_2
$1 \to 1$	$1 \to 3$	$1 \to 2$	$1 \to 4$
$2 \to 2$	$2 \to 4$	$2 \to 1$	$2 \to 3$
$3 \to 3$	$3 \to 1$	$3 \to 4$	$3 \to 2$
$4 \to 4$	$4 \to 2$	$4 \to 3$	$4 \to 1$

Symmetries of a square

1_S	r_1	r_2	r_3
$1 \to 1$	$1 \to 2$	$1 \to 3$	$1 \to 4$
$2 \to 2$	$2 \to 3$	$2 \to 4$	$2 \to 1$
$3 \to 3$	$3 \to 4$	$3 \to 1$	$3 \to 2$
$4 \to 4$	$4 \to 1$	$4 \to 2$	$4 \to 3$

m_1	m_2	d_1	d_2
$1 \to 2$	$1 \to 4$	$1 \to 3$	$1 \to 1$
$2 \to 1$	$2 \to 3$	$2 \to 2$	$2 \to 4$
$3 \to 4$	$3 \to 2$	$3 \to 1$	$3 \to 3$
$4 \to 3$	$4 \to 1$	$4 \to 4$	$4 \to 2$

Now suppose we want to flip and then rotate the rectangle. This turns out to be the same as applying two functions to $\{1, 2, 3, 4\}$, one after another. That is, we have a geometric interpretation of function composition. For example, let's flip the rectangle using the motion m_1 and then rotate by the motion r_1. We can use the arrow notation to record what happens to the vertices, and denote the resulting composition by $r_1 \circ m_1$ ("m_1 followed by r_1"):

$$1 \rightarrow 2 \rightarrow 4$$
$$2 \rightarrow 1 \rightarrow 3$$
$$3 \rightarrow 4 \rightarrow 2$$
$$4 \rightarrow 3 \rightarrow 1.$$

This composition is the same as the single motion described by m_2. In group theory, we record this information in a table, called a **Cayley table**. The answer, m_2, is entered in the row labeled by r_1 and the column headed by m_1. (Note that some textbooks use an alternate convention in recording the compositions in the table. Be sure you know the convention your text uses.)

7. Complete the Cayley tables for the triangle, square and rectangle.

Symmetries of a rectangle

$*$	1_R	r_1	m_1	m_2
1_R				
r_1			m_2	
m_1				
m_2				

Symmetries of a triangle

*	1_T	r_1	r_2	m_1	m_2	m_3
1_T						
r_1						
r_2						
m_1						
m_2						
m_3						

Symmetries of a square

*	1_S	r_1	r_2	r_3	m_1	m_2	d_1	d_2
1_S								
r_1								
r_2								
r_3								
m_1								
m_2								
d_1								
d_2								

8. Did it matter in what order you did the motions? Explain.

1.2   In the Lab

The **order** of a group G is the number of elements in G. *Exploring Small Groups (ESG)* has a **Group Library** (see Appendix C), which contains all groups with orders between 3 and 16. The important part of the tables you constructed is the pattern exhibited–the names of the permutations don't really matter. In this part of the lab, you'll try to match the tables you constructed with the ones stored in the computer.

Let's start with the symmetries of the rectangle. You constructed four different permutations. So we have to investigate the groups of order 4 in the computer library. Choose option 4 (Operation Table from the Group Library) from the **Table Generation Menu.** Notice that *ESG* lists two groups of order 4, numbers 0401 and 0402, listed in the **Group Library**. Type in "0401." You should see a table displayed.

9. Does the table on the computer screen have the same pattern as the one you constructed? This may be hard to see since the elements in group 0401 have numbers as names. Using option 7 (Table Alterations), try to rename the elements so that they agree with your names. You may have to do this several times to see how the table works. You may also want to try to reorder the elements. Can you make group 0401 look like your table?

10. Do the same thing for group 0402. When you finally have a match of your table and the table on the computer, record how you changed and rearranged the names of the elements. Make a note of which number the computer uses for the group of symmetries of a rectangle.

11. You can see in the **Group Library** that group 0602 is the group of symmetries of the triangle. Display that table. Rename the elements, and reorder if necessary, so that you have duplicated the table in your lab. Be sure to record how you renamed and reordered the elements.

12. Display the other group of order 6 (0601). Explain why the pattern of this group cannot be the same as the group of symmetries of a triangle, no matter how you rename and reorder the elements.

13. Repeat questions 11 and 12 for the symmetries of a square, using group 0804. As before, be sure to record how you renamed and reordered the elements of the group in the computer.

14. Choose another group of order 8 and explain why this group is different from the group of symmetries of a square.

1.3 Further Work

15. Based on the work you did with the rectangle, the triangle and the square, complete the following sentences:

    (a) A rotation followed by a rotation is always a _____.

    (b) A flip followed by a flip is always a _____.

    (c) A flip followed by itself is always the _____.

    (d) A flip followed by a rotation is always a _____.

    (e) A rotation followed by a flip is always a _____.

16. Think about the regular pentagon (5 sides) and the regular hexagon (6 sides). Give geometric descriptions of the symmetries of each shape. How many rotations would you expect to have? How many flips?

17. Generalize your observations to a regular n-gon (n sides). Give geometric descriptions of the symmetries. You'll need to consider two cases: n even and n odd. Would you expect the statements you made in problem 15 to hold for the general case? Why or why not?

1.4 Geometric Shapes

Make one copy of each shape on the next page to use in the activities of the lab.

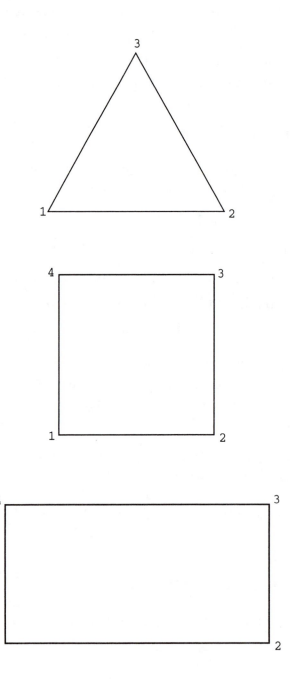

Chapter 2

Cayley Tables

We often find it convenient to present information about the **binary operation** of a group in a table, called a **Cayley table**. In this lab, you'll see how these tables are constructed and how we can identify which tables characterize groups.

2.1 Binary Operations on Sets

2.1.1 Before the Lab

Definition 2.1: Let S be a set. A **binary operation** $*$ on S is a rule which assigns an element $c \in S$ to each ordered pair (a, b) of elements of S. We write $a * b = c$.

Alternately, we say that S is **closed** under an operation $*$ if $a * b \in S$ for all $a, b \in S$.

Definition 2.2: A set G, together with an operation $*$, is called a **group** if the following conditions hold:

      i. The operation $*$ is a binary operation on G;

      ii. The operation $*$ is associative;

      iii. There exists an element $1_G \in G$, called the **identity element**, such that $1_G * a = a * 1_G$ for all $a \in G$;

iv. For every $b \in G$, there exists an element $b^{-1} \in G$, called the **inverse** to b, such that $b * b^{-1} = b^{-1} * b = 1_G$.

We will be able to use *ESG* to check whether a random or predetermined binary operation defined on a set satisfies these group properties.

2.1.2   In the Lab

1. Choose option 2 (Random Operation Table) from the **Table Generation Menu** to construct three operation tables, each defined on a set with four elements. Record each table. For each example, *ESG* will give you the answers to the following questions:

   (a) Does the table define an associative operation?

   (b) Is there a two-sided identity?

   (c) Does each element have an inverse?

2. Use option 3 (User-Defined Operation Table) to define a binary operation on a set of 4 elements. (Don't work too hard on this— we will use a more sophisticated method to construct a chart in section 2). Be sure to record your table. Check the group properties using option 6—did any fail? Did you define a group?

Now we will pick a table already defined by *ESG*. Select option 1 (Operation Table from the Sample Library) from the **Table Generation Menu** and choose number 1. Here, the set is

$$\{0, 1, 2, 3, 4, 5, 6, 7, 8\}.$$

with operation $*$ defined as multiplication mod 9. For example, $2 * 5 = 1$, because $2 \cdot 5 = 10$, and $10 \equiv 1$ mod 9.

3. Use *ESG* to check associativity, identity and inverses. Be sure to monitor the computations. Is this example a group? Why or why not?

ESG tells you whether the operation is a group operation. By now, you should be convinced that most binary operations defined on a set S will not define a group operation on S. For example, there are 3^9 ways to complete the table

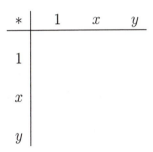

to give a binary operation on the set $\{1, x, y\}$ (why?), but only one will satisfy all the group properties. In section 2, we'll use two basic facts about groups to streamline the construction of possible tables.

2.2   Cayley Tables for Groups of Orders 2, 3 and 4

2.2.1   Before the Lab

Definition 2.3: We say that an operation table for a binary operation is a **Cayley Table** if it defines the binary operation for a group.

Now suppose that G is a group, with binary operation $*$. We'll need to use the following two facts about groups:

i. **Identity Property**: The identity element $1_G \in G$ is defined by the equation:

$$a * 1_G = 1_G * a = a \text{ for every } a \in G$$

ii. **Unique Solutions Property**: Let a and b be elements of G. The equations $a * u = b$ and $v * a = b$ have unique solutions in G.

The identity property is part of the definition of a group, and the cancellation property can be easily proved from the definition.

4. Carefully explain in your own words why the unique solutions property implies that each element of G must appear exactly once in each row and each column of the operation table.

5. Use these two properties to construct all possible Cayley tables for the set $S = \{1, x\}$ and the set $T = \{1, x, y\}$.

Now we will use these two properties to construct all possible Cayley tables for the set $W = \{1, x, y, z\}$. We start by filling in the first row and column using the identity property:

*	1	x	y	z
1	1	x	y	z
x	x	?		
y	y		??	
z	z			

We can replace ? by the elements 1, y or z.

6. Choose y and complete the chart using property ii.

7. Choose 1 and complete the chart in two ways, depending on your choice for ??.

8. Explain why replacing ? with z will yield results equivalent to those obtained when we replaced ? with y.

2.2.2   In the Lab

9. Choose option 3 (User-Defined Operation Table) from the **Table Generation Menu**, and enter each group you obtained in your answers to problems 5, 6, and 7. Confirm that you have defined a group in each case.

You should have obtained only one group of order 3 with your answer to problem 5. If you considered all the possibilities for tables for the set W, you should have constructed more than two tables. But *ESG*'s library of groups only lists two groups of order four.

10. Try to identify each of the groups you have discovered with one of the two listed in the *ESG* library. (Note: This may take some work. In some cases, you may have to rename and reorder the elements of the group on the computer, using option 7 from the **Group Properties Menu**.)

When we can match up the patterns of the Cayley tables of two groups by renaming and reordering the elements of one of the groups, we say the two groups are **isomorphic**.

11. List some characterizing properties of these Cayley tables so that you could make an easy identification of a group of order 4. For example, which elements appear along the diagonal of the table in each case?

Chapter 3

Cyclic Groups and Cyclic Subgroups

The most elementary groups are those that can be constructed from a single element. These are called **cyclic groups**. You will investigate them by hand and with *ESG*, and also see how they are contained inside more complicated groups.

3.1 Before the Lab

We define Z_8 to be the group formed by the set $\{0, 1, 2, 3, 4, 5, 6, 7\}$ with the binary operation $*$, addition mod 8. If we consider the element $2 \in Z_8$, we can compute

$$2 * 2 = 4$$
$$2 * 2 * 2 = 6$$
$$2 * 2 * 2 * 2 = 0.$$

That is, the element 2 generates the set $S = \{0, 2, 4, 6\}$.

1. Show explicitly that the set S is closed under the operation $*$ and that the inverse of every element in S is also in S.

   Since $0 \in S$ and $*$ is associative, we see that S is a group as well as a subset of Z_8. We say that S is a **subgroup** of Z_8, and that the element

15

2 generates the subgroup S. We write $Sg(2)$ to denote the subgroup generated by the element 2.

Definition 3.1: Let G be a group and H a subset of the set G. If H is a group under the binary operation of G, we call H a **subgroup** of G.

2. Compute other subgroups of Z_8 by following the same procedure for each element of Z_8. Record your data in a chart with the headings:

$$\text{element } x \in Z_8 \qquad Sg(x)$$

Notice that several elements generate the whole group Z_8. When this happens, we say the group is **cyclic** and call each element x that gives us the whole group a **generator** of the group.

Let's try this again with S_3. As before, we calculate the powers of an element to generate a cyclic subgroup. For example, if we choose $r_1 \in S_3$, we have

$$r_1^1 = r_1$$
$$r_1^2 = r_1 * r_1 = r_2$$
$$r_1^3 = r_1 * r_1 * r_1 = 1_T.$$

Thus, $Sg(r_1) = \{1, r_1, r_2\}$.

3. Generate all the subgroups you can for S_3 by repeating this procedure for all the other elements of S_3. Record your answers in a chart as before, with headings:

$$\text{element } x \in S_3 \qquad Sg(x)$$

4. Is there any element that gives you the entire group in this case?

3.2 In the Lab

In the **Group Table Library** of *ESG*, it is easy to identify all the cyclic groups. In each case, the cyclic group of order n is listed first among all the groups of that order. *ESG* uses the notation Cn; here we will use the notation Z_n. Choose the group Z_{12} (see Appendix 3 for the four-digit code) from the **Group Library**, and select option 1 (Powers and Orders) from the **Group Properties Menu**.. Enter the element 2 in answer to the question "For which element would you like to see powers and order calculated?" and "Y" to the question "Would you like to step through each of the calculations?" *ESG* uses *multiplicative* notation for all groups. So, for example, what you are used to writing as $2+2+2$ will be written as $2\char94 3$. The computer tells us that the order of 2 is 6 (that is, $2\char94 6 = 0$, or, in the usual notation, $2+2+2+2+2+2 \equiv 0$ *mod* 12.)

Definition 3.2: Let G be a group and $x \in G$. The **order** of the element x is the smallest positive integer r so that $x\char94 r = 1_G$.

The computer also tells us that $Sg(2) = \{0, 2, 4, 6, 8, 10\}$. This means the subgroup formed by all powers of the element 2 is $Sg(2)$.

5. Repeat this procedure for all elements of Z_{12}. Then record the information you obtain in a chart with the headings:

   $$\text{element } x \in Z_{12} \qquad Sg(x) \qquad \text{order of } x$$

   What is the relationship between the second and third columns of the chart?

   Notice that for several of the elements $x \in Z_{12}$, we have that $Sg(x) = Z_{12}$.

6. Use option 1 (Powers and Orders) in the **Group Properties Menu** to obtain $Sg(x)$ and the order of x for every element $x \in Z_5$; repeat for Z_7. Create a chart as above for each group.

7. It should now be clear to you that Z_n is *always* a cyclic group. Based on the data you have obtained for Z_5, Z_7, Z_8, and Z_{12}, make a conjecture about the possible generators of Z_n. (You may want to consider several cases for values of n.)

8. Use option 1 (Powers and Orders) in the **Group Properties Menu** to find the orders of all elements and to generate all cyclic subgroups for the group D_4. Record the data for each element in the chart below, using the names of the elements as they appear on the computer. First confirm that your computer gives the same results as those already entered in the chart.

element $x \in D_4$	$Sg(x)$	order of x
A	$\{1, A, B, C\}$	4
D	$\{1, D\}$	2

9. Explain why D_4 is not cyclic.

10. Use option 1 (Powers and Orders) in the **Group Properties Menu** to find the orders of all elements and to generate all cyclic subgroups (that is $Sg(x)$, for all $x \in G$) of the groups listed below. Use Appendix 3 to determine the four-digit code for each group.

$$\begin{array}{ll} Q_4 & D_7 \\ D_5 & Z_8 \times Z_2 \\ D_6 & M \\ A_4 & D_8 \\ Q_6 & Q_8 \end{array}$$

Collect your information in a chart for each group G with the headings:

$$\text{element } x \in G \qquad Sg(x) \qquad \text{order of } x$$

Be sure to use the names of the elements of each group that are used by *ESG*. You'll be using the subgroups again in Lab 4, so record your information carefully!

3.3 Further Work

To be complete, we include the following formal definitions of cyclic groups and subgroups.

Definition 3.3: Let G be a group and $a \in G$. We write

$$<a> = \{a^n \mid n \in Z\}$$

and say that $<a>$ is the **cyclic subgroup** generated by a.

Definition 3.4: We say that G is **cyclic** if $G = <a>$ for some $a \in G$. We call a the **generator** of the group G.

11. In definition 3.3, there are an infinite number of powers n. *ESG* only has finite groups in its library. Can this definition work for *finite* cyclic groups? Explain your answer carefully.

12. Show that the set of integers, Z, under the binary operation of addition is a cyclic group.

13. Prove that a group with no proper nontrivial subgroups is cyclic.

14. Suppose that G is a group that is not cyclic. Make a conjecture about the orders of the elements of G.

Chapter 4

Subgroups and Subgroup Lattices

A **subgroup lattice** is a convenient and informative way of presenting the subgroups of a group. You will use *ESG* to generate the subgroups of a collection of groups. Then, you will arrange the subgroups into diagrams called lattices. Throughout the semester you will use these subgroup lattices to provide information about the groups you study. The activities in this lab will, therefore, prepare important reference material for the rest of the course.

4.1   Before the Lab

ESG will do most of the computational work for us in this lab. But before we turn to the computer, we need to work out a few examples by hand. We start with the Klein four-group, V, the symmetries of the triangle, S_3, and a cyclic group, Z_8. You have already found the subgroups of these groups, in Lab 3.

group	subgroups
V	$\{1_R\}, \{1_R, r\}, \{1_R, d_1\}, \{1_R, d_2\}, V$
S_3	$\{1_T\}, \{1_T, m_1\}, \{1_T, m_2\}, \{1_T, m_3\}, \{1_T, r_1, r_2\}, S_3$
Z_8	$\{0\}, \{0, 4\}, \{0, 2, 4, 6\}, Z_8$

In a subgroup lattice, the full group appears at the top and the trivial subgroup at the bottom. Intermediate subgroups are arranged by order, with connecting lines showing containment. Thus, the lattice for Z_8 is

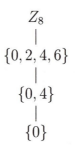

But the proper nontrivial subgroups of V and S_3 are not contained in one another. The lattice for V is

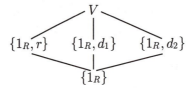

and the lattice for S_3 is

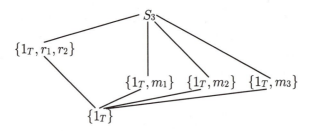

Notice that the three proper nontrivial subgroups of V are all on the same level, because each one has order 2. But the proper nontrivial subgroups of S_3 are on two different levels—one for those of order 2 and one for the subgroup of order 3. Be sure to make clean copy of each lattice in your notebook, *each on a separate page,* for future reference. You will want to record other relevant information on this sheet in subsequent labs.

4.2   In the Lab

Work through the following example using *ESG* before you try to do the problems. We will use the names of the elements of D_4 as they appear on the computer. So $D_4 = \{1, A, B, C, D, E, F, G\}$. In Lab 3, you found the cyclic subgroups of D_4 to be $\{1\}$, $\{1, B\}, \{1, D\}, \{1, E\}$, $\{1, F\}$, $\{1, G\}$, and $\{1, A, B, C\}$. Now, let's use the feature that allows us to add another generator to a subgroup. Choose option 3 (Subgroups and Cosets/Quotients) from the **Group Properties Menu**, and then routine 1 (Subgroup generated from elements you choose). First, start with the element B and generate the subgroup of order 2 with that. When *ESG* asks you to add another generator, add D. You should then obtain the subgroup $\{1, B, D, F\}$. What happens if you choose D and then F? A and then B?

1. Continue to generate subgroups with all possible pairs of generators. Record the information you obtained in a chart with the headings:

   pairs of elements   subgroup generated

   Confirm that the distinct subgroups you obtained are $\{1\}$, $\{1, B\}$, $\{1, D\}$, $\{1, E\}$, $\{1, F\}$, $\{1, G\}$, $\{1, A, B, C\}$, $\{1, B, D, F\}$, $\{1, B, E, G\}$ and D_4.

2. Are there any subgroups of D_4 that might need three generators? How many generators do you need to obtain the full group?

3. What would be the most efficient way to obtain all possible subgroups of a group using *ESG*? In a sentence or two, describe your method.

The subgroups of D_4 can be assembled into the following lattice:

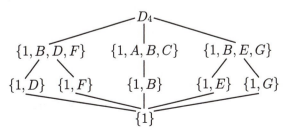

Confirm that this agrees with the data you obtained with *ESG*.

4. Construct the subgroup lattice for each of the groups listed below. For each one, use the names of the elements as stated in the computer program. Draw each lattice carefully on a separate sheet. You might want to print out the Cayley table for each group for future reference.

$$
\begin{array}{ll}
Q_4 & D_7 \\
D_5 & Z_8 \times Z_2 \\
D_6 & M \\
A_4 & D_8 \\
Q_6 & Q_8
\end{array}
$$

5. Try to identify the subgroups you found as familiar groups (e.g., V or Z_3). Indicate these identifications on your lattice.

6. Consider the group D_n, the symmetries of a regular n-gon.

   (a) Discuss how subgroups can be constructed using the geometry you developed in Lab 1. Identify at least one nontrivial subgroup in D_n, $n > 3$, which is not cyclic. You may need to consider the cases n even and n odd separately. Make a guess at what might happen when n is odd but not prime.

   (b) Argue geometrically that for any n, D_n has a subgroup having exactly half of the elements of the group.

4.3   Further Work

7. If G is any group of order n, make a conjecture about the order of any subgroup H of G.

8. If $G = D_n$, of order $2n$, make a conjecture (different from the one you made in problem 7) about the subgroups of G and their orders.

9. Write out the converse of the conjecture you wrote for problem 7. Could it possibly be true in general? Are there groups in the lab for which it could be true?

10. What does it mean for two groups to have the same subgroup lattice?

11. Would it be possible for two groups of the same order to have the same subgroup lattice but not be isomorphic? Explain your answer.

12. You do not have a lot of examples from which to draw conclusions. How confident do you feel about your answers to problems 6-11? If you had a way to look at additional groups, which would you choose to test your conjectures?

Chapter 5

The Center and Commutator Subgroups

If a group G is not abelian, you might want to ask, "How close is G to being abelian?" Two special subgroups, the center and the commutator subgroup, can provide some answers.

5.1   Before the Lab

Definition 5.1: For any group G, let

$$Z(G) = \{x \in G \mid xa = ax \text{ for all } a \in G\}.$$

We call $Z(G)$ the **center** of G.

For homework, you will prove that $Z(G)$ is a subgroup of G.

Definition 5.2: Let G be a group. The **commutator subgroup** G' of G is the subgroup generated by the elements of the form $aba^{-1}b^{-1}$, with $a, b \in G$.

For some groups, you will need to add additional elements to the set to form a subgroup after you have computed all possible $aba^{-1}b^{-1}$.

Example: We will work through one example by hand, computing the center and commutator subgroup of S_3.

Center of S_3:

element	relation	conclusion
1_T	$1_T x = x 1_T \ \forall x \in G$	$1_T \in Z(S_3)$
r_1	$r_1 m_1 \neq m_1 r_1$	$r_1 \notin Z(S_3)$
r_2	$r_2 m_1 \neq m_1 r_2$	$r_2 \notin Z(S_3)$
m_1	$m_1 m_2 \neq m_2 m_1$	$m_1 \notin Z(S_3)$
m_2	$m_2 m_1 \neq m_1 m_2$	$m_2 \notin Z(S_3)$
m_3	$m_3 m_1 \neq m_1 m_3$	$m_3 \notin Z(S_3)$

Therefore, $Z(S_3) = \{1_T\}$.

Commutator subgroup of S_3:

elements $a,\ b$	commutator $aba^{-1}b^{-1}$
r_1, r_2	$r_1 r_2 r_2 r_1 = 1_T$
$m_1,\ r_1$	$m_1 r_1 m_1 r_2 = r_1$

1. Complete the table for the commutator subgroup of S_3. Do the elements you have calculated form a subgroup of S_3?

5.2   In the Lab

2. Use the option 3 (Subgroups and Cosets/Quotients) of the **Group Properties Menu** to complete the chart below:

group	center $Z(G)$	commutator subgroup G'
V		
Z_5		
S_3		
Z_8		
D_4		
Q_4		
D_5		
D_6		
A_4		
Q_6		
D_7		
M		
D_8		
Q_8		

5.3 Further Work

3. Write a statement which characterizes the center of an abelian group.

4. Write a statement which characterizes the commutator subgroup of an abelian group.

5. Make a conjecture about the center of D_n. (Hint: you may have to consider several cases.)

6. Make a conjecture about the commutator subgroup of D_n.

7. Record the center and commutator subgroup on each subgroup lattice in your notebook. Use a consistent code—for example, you may want to underline the center in blue or circle each center.

8. Let M_2 be the set of all 2×2 invertible matrices with entries in the real numbers R. This forms a group under matrix multiplication. What is the center of M_2? Make a conjecture about the center of M_n, the group of $n \times n$ invertible matrices with entries in R.

9. Prove that $Z(G)$ is a subgroup of G, for any group G.

10. In your own words, explain how the center and commutator subgroup of a group G give an answer to: "How close is G to being abelian?"

Chapter 6

Quotient Groups

The elements of a group G can be partitioned, using a subgroup H of G, into cells called **cosets**. One of the nicest features of ESG will allow you to investigate these cosets, which sometimes form a new group called a **factor** or **quotient group** of G.

6.1 Before the Lab

You will prove the following theorem for homework:

Theorem 6.1: Let G be a group and H a subgroup of G. Define $\sim$ by saying $a \sim g$ if and only if $ag^{-1} \in H$. Then $\sim$ is an equivalence relation.

We would obtain a similar result if we defined $\sim$ by saying $a \sim g$ if and only if $g^{-1}a \in H$.

In the the theorem, we have $a = hg$, for some $h \in H$. In fact, as we let h run through all the elements of H, we obtain all elements of G that are related to g under the equivalence relation $\sim$. The cell of the partition created by $\sim$ that contains g is denoted by Hg and is called the **right coset** of H in G that contains g. In the second case, we create gH, the **left coset** of H in G that contains g. Why do both Hg and gH contain g?

For the two examples below, we'll consider the subgroups $H = \{1, r_1, r_2\}$ and $K = \{1, m_1\}$ of S_3. We'll compute the right and left cosets for each subgroup. Confirm the computations as you read through the example.

Example 1:

H : right cosets

$$\{1, r_1, r_2\}1 = \{1, r_1, r_2\}$$
$$\{1, r_1, r_2\}r_1 = \{1r_1, r_1r_1, r_2r_1\} = \{r_1, r_2, 1\}$$
$$\{1, r_1, r_2\}r_2 = \{1r_2, r_1r_2, r_2r_2\} = \{r_2, 1, r_1\}$$
$$\{1, r_1, r_2\}m_1 = \{1m_1, r_1m_1, r_2m_1\} = \{m_1, m_3, m_2\}$$
$$\{1, r_1, r_2\}m_2 = \{1m_2, r_1m_2, r_2m_2\} = \{m_2, m_1, m_3\}$$
$$\{1, r_1, r_2\}m_3 = \{1m_3, r_1m_3, r_2m_3\} = \{m_3, m_2, m_1\}$$

There are two distinct right cosets, $\{1, r_1, r_2\}$ and $\{m_1, m_2, m_3\}$, which form a partition of the elements of G.

H : left cosets

$$1\{1, r_1, r_2\} = \{1, r_1, r_2\}$$
$$r_1\{1, r_1, r_2\} = \{r_11, r_1r_1, r_1r_2\} = \{r_1, r_2, 1\}$$
$$r_2\{1, r_1, r_2\} = \{r_21, r_2r_1, r_2r_2\} = \{r_2, 1, r_1\}$$
$$m_1\{1, r_1, r_2\} = \{m_11, m_1r_1, m_1r_2\} = \{m_1, m_2, m_3\}$$
$$m_2\{1, r_1, r_2\} = \{m_21, m_2r_1, m_2r_2\} = \{m_2, m_3, m_1\}$$
$$m_3\{1, r_1, r_2\} = \{m_31, m_3r_1, m_3r_2\} = \{m_3, m_1, m_2\}$$

There are two distinct left cosets, $\{1, r_1, r_2\}$ and $\{m_1, m_2, m_3\}$. These agree with the right cosets: $Hg = gH$ for every $g \in G$.

Example 2:

K : right cosets	K : left cosets
$\{1, m_1\}1 = \{1, m_1\}$	$1\{1, m_1\} = \{1, m_1\}$
$\{1, m_1\}r_1 = \{r_1, m_2\}$	$r_1\{1, m_1\} = \{r_1, m_3\}$
$\{1, m_1\}r_2 = \{r_2, m_3\}$	$r_2\{1, m_1\} = \{r_2, m_2\}$
$\{1, m_1\}m_1 = \{m_1, 1\}$	$m_1\{1, m_1\} = \{m_1, .1\}$
$\{1, m_1\}m_2 = \{m_2, r_1\}$	$m_2\{1, m_1\} = \{m_2, r_2\}$
$\{1, m_1\}m_3 = \{m_3, r_2\}$	$m_3\{1, m_1\} = \{m_3, r_1\}$

There are three distinct right cosets, $\{1, m_1\}$, $\{r_1, m_2\}$, and $\{r_2, m_3\}$. There are three distinct left cosets, $\{1, m_1\}$, $\{r_1, m_3\}$, and $\{r_2, m_2\}$. This time, however, not all of the left and right cosets agree. That is, there are elements $g \in S_3$ so that $Kg \neq gK$. The distinction between H and K is crucial and at the heart of studying quotient groups. The collection of cosets forms a new group precisely when the left and right cosets agree. The problem, it turns out, is in trying to define a binary operation on the cosets.

The **coset operation** is best defined through an example. We only have the operation defined on G to use. Let's start by using the left cosets of H. Set $X = \{1, r_1, r_2\}$ and $Y = \{m_1, m_2, m_3\}$. Choose an element from X and one from Y, for example, $r_1 \in X$ and $m_2 \in Y$. Since $r_1 m_2 = m_1$, and $m_1 \in Y$, we define the coset operation $*$ by $X * Y = Y$. No matter which representative of the cosets X and Y we choose, we will obtain the same answer in this case. So we can construct the Cayley table using the operation $*$:

$*$	X	Y
X	X	Y
Y	Y	X

Check that this forms a group of order 2. Which coset serves as the identity element?

If we try to do this with the left cosets of K, however, we run into problems. Set $U = \{1, m_1\}$, $V = \{r_1, m_3\}$, $W = \{r_2, m_2\}$. If we compute $V * W$ using the elements m_3 and r_2, we have $V * W = W$. But if we do the calculation with r_1 and m_2, we obtain $V * W = U$. Obviously there is a problem here—the operation is not well-defined. In the lab, we will see how ESG demonstrates this with color.

To summarize, we have the theorem:

Theorem 6.2: Let G be a group and let H be a subgroup of G. The cosets of H in G form a group if and only if $Hg = gH$ for all $g \in G$.

When the condition of the theorem holds, we will say that the subgroup H is a **normal** subgroup of G. The collection of cosets (right or left, of course) is then called the **quotient** or **factor group**, G mod H, denoted by G/H. In our examples above, the subgroup H is normal but K is not.

It is especially important for you to work out one example carefully on your own before you use the computer. Do all the computations **by hand** for question 1.

1. Consider the subgroups $H_1 = \{1, r_2, d_1, d_2\}$, $H_2 = \{1, m_1\}$, and $H_3 = \{1, r_2\}$ of D_4.

   (a) Find all the left cosets of the subgroups in D_4.

   (b) Find all the right cosets of the subgroups in D_4.

   (c) For which subgroups are the left and right cosets equal?

   (d) For each subgroup identified in part (c), construct a group table for the quotient group G/H_i. What familiar group has the same group table?

Be sure to bring your subgroup lattices to the lab.

6.2   In the Lab

Check your answers to question 1 with *ESG* before continuing with the problems below. Choose option 3 (Subgroups and Cosets/Quotients) from the **Group Properties Menu** for D_4, and generate each subgroup H_i. Answer "Y" to the question, "Would you like to see the left cosets of this subgroup?" Look at the coloring of the Cayley table of D_4, grouped by the left cosets of each subgroup. In some cases, you will be asked, "Would you like to see the quotient table?" Be sure you understand how the Cayley table is transformed when you answer "Y."

2. In your own words, explain how you can determine from the table on the computer screen that the coset operation is well-defined or not well-defined.

For problems 3-10, you may use the computer for your computations. Answer the following questions for each group G and **all** non-trivial proper subgroups H of G.

(a) Record the **distinct** left and right cosets of the subgroup H in G.

(b) Is the subgroup H normal in G? If so, have the computer construct a group table for the new quotient group G/H. What familiar group has the same group table? Your answer should be a known group from the *ESG* library. Be sure to record on your subgroup lattices which subgroups are normal.

Note that there are no general formulas which help us figure out which group G/H actually is. You have to rely on computing the order of G/H and on knowing something about the various groups of that order.

3. $G = D_5$

4. $G = D_6$

5. $G = A_4$

6. $G = Q_6$

7. $G = D_7$

8. $G = M$

9. $G = D_8$

10. $G = Q_8$

6.3 Further Work

11. Make at least two conjectures about the kinds of subgroups which always seem to be normal in a finite group G. (Hint: think about the special subgroups that we considered in an earlier lab or the index of the subgroup).

12. Make at least two conjectures about the factor groups D_n/H, where H is either the commutator subgroup or center of D_n.

13. Prove Theorem 6.1: Let G be a group and H a subgroup of G. Define $\sim$ by $a \sim g$ if and only if $ag^{-1} \in H$. Then $\sim$ is an equivalence relation.

Chapter 7

Direct Products

Familiar and elementary groups can be used as building blocks to create more complicated groups. In this lab, you'll analyze the structure of a few groups and construct some new ones using **direct products**.

7.1 Before the Lab

7.1.1 External Direct Products

Definition 7.1: Let G and K be two groups. Define $G \times K$ by

$$G \times K = \{(g, k) \mid g \in G, k \in K\}$$

with the operation $*$ defined on $G \times K$ by

$$(g_1, k_1) * (g_2, k_2) = (g_1 g_2, k_1 k_2), \text{ for } g_1, g_2 \in G, k_1, k_2 \in K,$$

where the operation in the first coordinate is the operation of G and the operation in the second coordinate is the operation of K. We say that $G \times K$ is the **external direct product** of G and K.

1. Prove that this construction gives us a group.

2. What is the order of $G \times K$ when G and K are finite groups?

If $G = Z_2$ and $K = Z_2$, then $G \times K = \{(0,0), (1,0), (0,1), (1,1)\}$. It is not hard to check that the nonidentity elements all have order two. There is only one noncyclic group of order 4, the Klein four-group, V. Hence $G \times K \simeq V$.

3. For G and K specified below, write out the elements of $G \times K$ and determine the order of each element. Which of the direct products are cyclic?

   (a) $G = Z_4$, $K = Z_2$

   (b) $G = Z_3$, $K = Z_2$

   (c) $G = Z_6$, $K = Z_2$

4. Make a conjecture about the structure of $G \times K$ when G and K are both cyclic groups.

7.1.2   Internal Direct Products

Now suppose that H_1 and H_2 be subgroups of a group G. We can certainly form the external direct product $H_1 \times H_2$. But the elements of $H_1 \times H_2$ are no longer elements of the group G. So we ask, is there some way to generate a new subgroup of G using H_1 and H_2?

We start by defining the *subset* $H_1 H_2$ of G,

$$H_1 H_2 = \{xy \mid x \in H_1, y \in H_2\}$$

If S is a set, let $\#(S)$ denote the number of elements in S.

5. What is the maximum number of elements possible in $H_1 H_2$? Think of some reasons why $\#(H_1 H_2)$ can be less than the maximum.

6. What is a necessary condition in order that $\#(H_1 H_2) = \mid H_1 \times H_2 \mid$?

For homework, you will prove that $H_1 H_2$ is a subgroup of G when G is abelian. But when G is nonabelian, $H_1 H_2$ may or may not be a subgroup. We can see this in the example of D_8. Consider the subgroups

$$H_1 = \{1, D\} \qquad H_2 = \{1, L\} \qquad H_3 = \{1, K\} \qquad H_4 = \{1, M, I, D\}.$$

Then we can calculate some of the products $H_i H_j$, $1 \leq i, j \leq 4$. For example, we have

$$H_1 H_4 = \{1 \cdot 1, 1M, 1I, 1D, D1, DM, DI, DD\} = \{1, M, I, D\}$$
$$H_2 H_3 = \{1 \cdot 1, 1K, L1, LK\} = \{1, K, L, A\}$$
$$H_1 H_2 = \{1 \cdot 1, 1L, D1, DL\} = \{1, D, L, H\}.$$

We see that $H_1 H_4$ and $H_1 H_2$ are subgroups of D_8, but $H_2 H_3$ is not.

7. Calculate all the other possible $H_i H_j$, $1 \leq i, j \leq 4$. Which are subgroups of D_8?

8. Let H_1 and H_2 be subgroups of an arbitrary group G. Make a preliminary conjecture about when $H_1 H_2$ is a subgroup of G based on your data.

We can also form the external direct product with these subgroups. For example,

$$H_1 \times H_2 = \{(1, 1), (1, L), (D, 1), (D, L)\}.$$

We note that $H_1 \times H_2$ is a group of order 4 which is not cyclic. Thus it must be isomorphic to V, the Klein four-group. But $H_1 H_2 = \{1, D, L, H\}$ is also isomorphic to V. That is, we have that $H_1 \times H_2 \simeq H_1 H_2$.

9. Determine the structure of $H_i \times H_j$, $1 \leq i, j \leq 4$. Decide for which pairs of subgroups it is true that $H_i \times H_j \simeq H_i H_j$.

We can summarize these ideas with the following formal definition:

Definition 7.2: Let H_1, H_2 and K be subgroups of a group G. We say that K is the **internal direct product** of H_1 and H_2 if $K \simeq H_1 \times H_2$.

10. Let H_1 and H_2 be subgroups of an arbitrary group G. Make a preliminary conjecture about when $H_1 \times H_2 \simeq H_1 H_2$.

7.2   In the Lab

11. Using *ESG*, construct complete, annotated subgroup lattices for groups 1608 and 1613. Include as much information as possible, such as which subgroups are normal and to which familiar groups the subgroups might be isomorphic.

12. Now consider the groups $G = 1608$ and 1613.

    (a) Find two subgroups H_1, H_2 in each G for which $H_1 H_2$ is a subgroup of G *and* $H_1 H_2 \simeq H_1 \times H_2$. Be sure to explain why the isomorphism holds.

    (b) Find two subgroups H_1, H_2 in each G for which $H_1 H_2$ is a subgroup of G *and* $H_1 H_2$ is not isomorphic to $H_1 \times H_2$. Explain why there is no isomorphism for your example.

7.3   Further Work

13. Let G be an abelian group and let H_1 and H_2 be subgroups. Prove that $H_1 H_2$ is a subgroup of G.

14. Suppose that G is an arbitrary group. Refine the conjecture you made in problem 8 about when $H_1 H_2$ can be a subgroup of a group G.

15. Refine the conjecture that completes the following statement: Let H_1 and H_2 be subgroups of a group G. Then $H_1 \times H_2 \simeq H_1 H_2$ if ____. (Hint: be sure that you include the conditions that will guarantee that $H_1 H_2$ is a subgroup of G.).

Chapter 8

The Unitary Groups

The collection of multiplicative units in Z_n, $U(n)$, forms a group under multiplication modulo n. In this lab, you will investigate the structure of the group $U(n)$ and also analyze the structure of some abelian groups.

8.1   Before the Lab

We know that $Z_8 = \{0, 1, 2, 3, 4, 5, 6, 7\}$ forms a group under addition modulo 8. Suppose, however, that we consider these elements under multiplication modulo 8. That is , for $a, b \in Z_8$, define $a * b = ab \ mod$ 8. Can we think of Z_8 as a group under this multiplication? Certainly the element 1 can serve as the identity. We see that $5 * 5 = 1$. But it is impossible to find an element x so that $2 * x = 1$. Why? To form a group under the binary operation $*$, we'll need to eliminate some of the elements in the set. Fill out the following table using the operation $*$:

$$
\begin{array}{c|cccc}
* & 1 & 3 & 5 & 7 \\
\hline
1 & & & & \\
3 & & & & \\
5 & & & & \\
7 & & & &
\end{array}
$$

The operation $*$ is clearly an associative binary operation and 1 serves as the identity. Use the table to identify the inverse of each of the elements. Thus, we have a group of order 4 which we will call $U(8)$. To which familiar group is it isomorphic?

Definition 8.1: We define $U(n)$ as the abelian group of multiplicative
units in Z_n, under the binary operation of multiplication modulo
n.

8.2   In the Lab

We will use hand calculations and *ESG* to try to determine the structure of $U(n)$ for every n. The author of the software has presented a lot
of information on the **F2** screen about the various groups listed under
both option 1 (Operation Table from the Sample Library) and option 4
(Operation Table from the Group Library) of the **Table Generation
Menu**. You are free to use any information stored in the program as
you gather the data for this lab.

We'll work through an example to give you an idea about how to
proceed with this lab. Under option 1 (Operation Table from the Sample Library), choose number 13, $U(21)$. First, check option 6 (Group
Properties); then move onto the **Group Properties Menu**. We see
that $U(21)$ is a group of order 12. There are, however, two abelian
groups of order 12, Z_{12} and $Z_6 \times Z_2$. We need to identify some properties that characterize each of these groups before we can determine to
which one $U(21)$ is isomorphic. It is easy to start by listing the orders
of the elements of each group. Check the **F2** screen to see the orders of
the elements. Note that Z_{12} has 4 elements of order 12 while $Z_6 \times Z_2$ and
$U(21)$ have none. Confirm that the orders of the elements in $Z_6 \times Z_2$
and $U(21)$ agree. Does this tell you enough to identify $U(21)$? Explain
your reasoning. Also notice that the **F2** screen for $U(21)$ tells us that
$U(21) \simeq U(7) \times U(3)$.

1. By hand, calculate $U(n)$, $2 \le n \le 7$.

2. Make a chart with the following headings:

$$U(n)$$
order
direct product of other $U(k)$'s
direct product of cyclic groups

Fill in the chart with the information you know about $U(n)$, $2 \leq n \leq 7$, and $U(21)$. It is possible that you may not be able to fill in all the information for each group.

3. Complete the chart for all the unitary groups found under option 1 (Operation Table from the Sample Library) of the **Table Generation Menu** (8-23).

4. Insert the following data into the chart: $U(25) \simeq Z_{20}$, $U(27) \simeq Z_{18}$, $U(81) \simeq Z_{54}$, $U(125) \simeq Z_{100}$.

8.3 Further Work

5. Make a conjecture about the structure of $U(p)$, when p is prime, in terms of familiar cyclic or abelian groups,

6. Describe $U(2^r)$ completely by giving the structure of $U(2)$, $U(2^2)$, and then $U(2^r)$, $r \geq 3$.

7. Determine the order of $U(st)$ in terms of the orders of $U(s)$ and $U(t)$ when s and t are relatively prime.

8. Give a description of $U(st)$ in terms of other unitary groups when s and t are relatively prime.

9. Describe $U(p^r)$, p an odd prime, in terms of the familiar abelian groups.

10. Now suppose that the prime power decomposition of n is

$$p_1^{r_1} p_2^{r_2} p_3^{r_3} \cdots p_k^{r_k}.$$

As a summary statement of the results above, write a general formula that gives the structure of $U(n)$ in terms of familiar cyclic or abelian groups.

11. Use your conjecture to calculate $U(720)$ and $U(2400)$.

Chapter 9

Composition Series

In this lab, you will investigate relationships among subgroups of a group G. You will learn about **composition series** and **composition factors**. In addition, this lab provides background material for several important theorems in group theory.

9.1   Before the Lab

Definition 9.1: A group $G \neq \{1_G\}$ is called **simple** if the only normal subgroups of G are $\{1_G\}$ and G.

For homework, you will prove:

Proposition 9.1: Every finite simple abelian group is isomorphic to Z_p, p a prime.

Definition 9.2: Let N_i, $i = 0, ..., k$, be a collection of subgroups of a group G. We say that

$$\{1_G\} = N_0 \subset N_1 \subset ... \subset N_k = G$$

is a **composition series** of the group G if N_i is normal in N_{i+1} for all i, $0 \leq i \leq k - 1$, and if N_{i+1}/N_i is a simple group. We call N_{i+1}/N_i a **composition factor** of the series.

Now consider the subgroup lattice of the group G (1613)

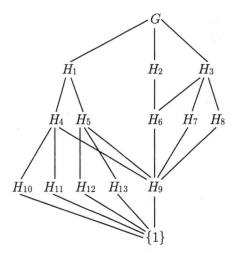

with subgroups

$H_1 = \{1, B, D, F, H, J, L, N\}$ $H_2 = \{1, A, B, C, D, E, F, G\}$
$H_3 = \{1, B, D, F, I, K, M, O\}$ $H_4 = \{1, D, H, L\}$
$H_5 = \{1, D, J, N\}$ $H_6 = \{1, B, D, F\}$
$H_7 = \{1, D, O, K\}$ $H_8 = \{1, D, I, M\}$
$H_9 = \{1, D\}$ $H_{10} = \{1, H\}$
$H_{11} = \{1, L\}$ $H_{12} = \{1, J\}$
$H_{13} = \{1, N\}$

Each of the following is a composition series for G

$$\{1\} \subset H_9 \subset H_4 \subset H_2 \subset G \tag{9.1}$$

$$\{1\} \subset H_{10} \subset H_5 \subset H_1 \subset G \tag{9.2}$$

$$\{1\} \subset H_9 \subset H_7 \subset H_3 \subset G \tag{9.3}$$

The composition factors for the composition series (9.1) are

$$H_9/\{1\} \simeq Z_2 \quad H_2/H_4 \simeq Z_2 \quad H_4/H_9 \simeq Z_2 \quad G/H_2 \simeq Z_2$$

where the isomorphisms hold because each subgroup has index two in the next subgroup of the series.

1. Compute the composition factors for (9.2) and (9.3). Then confirm that (9.2) and (9.3) are both composition series.

2. Find all additional composition series for G. In each case, be sure to compute the composition factors. Justify why each series is a composition series. (Hint: you may need to determine to which familiar group H_j is isomorphic, $1 \leq j \leq 13$).

3. Is $\{1\} \subset H_9 \subset H_4 \subset G$ a composition series? Explain why or why not.

4. Based on the data generated so far in this lab, make a conjecture about the composition series of a group.

This lab uses the subgroup lattices you have constructed in previous labs. It may be possible for you to complete the lab without any work on the computer at all. Or, you may find that you need to use the computer to construct a lattice or fill in details on a lattice you have previously constructed.

9.2   In the Lab

Be sure to bring the subgroup lattices for the groups Z_6, S_3, Z_7, D_4, Q_4, D_5, D_6, A_4, Q_6, D_7, M, D_8 and Q_8. If your lattices are incomplete for some of these groups, spend some time at the beginning of the lab constructing or completing them.

5. Construct all possible composition series for the groups listed above. Is there any group for which there is no composition series? Explain.

6. Compute the composition factors for each series you constructed in problem 5.

7. Suppose that G is a simple group. Write a composition series for G. What are the composition factors?

9.3   Further Work

8. Prove proposition 9.1.

9. Based on the data from this lab, make a conjecture about the composition factors of a composition series of a group G.

10. Refine the conjecture you made in problem 4 about the composition series of a group G.

11. Make a conjecture about the composition series and composition factors for Z_p, p a prime.

12. Make a conjecture about the composition series and composition factors for Z_n, n not a prime.

13. Make a conjecture about the composition series and composition factors for D_n, considering each of the following cases separately:

    (a) n prime

    (b) n odd, not prime

    (c) n even

Chapter 10

Introduction to Endomorphisms

This lab will introduce you to one type of function that is interesting to study in group theory, an **endomorphism**. It's especially nice to have the computer available for this topic—you'll be able to generate many examples without a lot of tedious calculations.

10.1   Before the Lab

Definition 10.1: Let G and K be groups. A function $T : G \to K$ is called a group **homomorphism** if $T(xy) = T(x)T(y)$ for every $x, y \in G$. When $G = K$, we say that T is a group **endomorphism**. An endomorphism that is one-to-one and onto is called an **automorphism**.

Because of the design of *ESG*, we will be looking at group endomorphisms and automorphisms rather than general homomorphisms.

Definition 10.2: The **kernel** of a homomorphism T is the set

$$\ker(T) = \{x \in G \mid T(x) = 1_G\}.$$

Definition 10.3: The **image** of a homomorphism T is the set

$$\mathrm{im}(T) = \{y \in K \mid \exists x \in G \text{ with } T(x) = y\}.$$

You will prove for homework that when T is an endomorphism these two sets are subgroups of the group G.

The computer will use the following two rules to compute the endomorphisms, once you choose some initial values for the function:

i. **Power rule:** If you set $T(a) = b$, then the computer calculates that $T(a^r) = b^r$.

ii. **Homomorphism rule:** If you set $T(a_1) = b_1$ and $T(a_2) = b_2$, then the computer calculates that $T(a_1a_2) = b_1b_2$.

1. Explain why the power rule is a special case of the homomorphism rule.

2. Using the power rule, explain why the order of b must divide the order of a, if $T(a) = b$.

3. Suppose that the group G is generated by the elements g_1 and g_2. Explain why choices for $T(g_1)$ and $T(g_2)$ will generate the full rule of the endomorphism T.

We will work through one example by hand and one on the computer before turning to the rest of the exercises. Consider the group S_3, which is generated by a rotation and a reflection (for example, r_1 and m_1). By the answer to problem 3, we know it will be sufficient to specify $T(r_1)$ and $T(m_1)$. Fill in the blanks as you read through the example.

Choose $T(r_1) = r_2$. The power rule then tells us that $T(r_1^2) = r_2^2$ or $T(r_2) = \_$, and $T(r_1^3) = r_2^3$ or $T(1_T) = \_$. Choose $T(m_1) = m_1$. Then $T(m_1^2) = m_1^2$, or $T(1_T) = \_$. Now use the homomorphism rule to expand the endomorphism. We have $T(m_1r_1) = T(m_1)T(r_1)$ or $T(m_2) = \_$. Also, $T(m_1r_2) = T(m_1)T(r_2)$, or $T(\_) = \_$. Confirm that all elements of S_3 have a value assigned by the rule of T. The computer can check easily that these values will or will not agree for all possible pairs, so we will complete this example in the lab. Have we defined an automorphism of S_3?

10.2   In the Lab

Be sure to work out the example below with *ESG* for the group of quaternions Q_4 before attempting to generate any endomorphisms on your own. Enter the values into the computer as you read through the example. First, using option 7 (Table Alterations) from the **Group Properties Menu**, rename the elements of Q_4 and reorder them to obtain the familiar table for the quaternions

$$
\begin{array}{ccccccccc}
1 & A & B & C & D & E & F & G \\
1 & i & -1 & -i & j & k & -j & -k
\end{array}
$$

It may also be helpful to write down the orders of the elements in the group: $i, -i, j, -j, k$, and $-k$ all have order 4, -1 has order 2 and of course the identity has order 1. Now choose option 5 (Endomorphisms) under the **Group Properties Menu**. We note that the elements i and j generate the group, so it will be sufficient to specify their images. For example, enter the element i and its image -1. This sets $T(i) = -1$. The computer will calculate

$$
\begin{aligned}
T(-1) &= T(i^2) = (-1)^2 = 1 \\
T(-i) &= T(i^3) = (-1)^3 = -1 \\
T(1) &= T(i^4) = (-1)^4 = 1.
\end{aligned}
$$

We need to consider several things before we specify $T(j)$. The element j is not contained in the subgroup $< i >$, but $j^2 = -1$ and we have already computed $T(-1)$ above. So any choice we make about $T(j)$ must agree with this. But -1 has order 2, so that will be acceptable as the image of j. Enter $T(j) = -1$. The computer will calculate

$$
\begin{aligned}
T(-1) &= T(j^2) = (-1)^2 = 1 \\
T(-j) &= T(j^3) = (-1)^3 = -1 \\
T(1) &= T(j^4) = (-1)^4 = 1.
\end{aligned}
$$

These computations agree with our previous work. We actually don't have any more choices to make, as the computer will use the homomorphism rule to compute the rest of the values. That is, we know that $k = ij$, so $T(k) = T(ij) = T(i)T(j) = (-1)(-1) = 1$. The power rule is used to compute the rest of the values

$$T(-1) = T(k^2) = 1^2 = 1$$
$$T(-k) = T(k^3) = 1^3 = 1$$
$$T(1) = T(k^4) = 1^4 = 1.$$

Once again this agrees with our previous work. The computer confirms that we have an endomorphism and we notice that $\ker(T) = \{1, k, -1, -k\}$ and $\text{im}(T) = \{1, -1\}$.

In this example, we have mapped elements of order 4 to an element of order 2. If we wanted to construct an automorphism, we would have needed to send each element of order 4 to an element of order 4 (why?). Verify using *ESG* that the following rule gives an automorphism of Q_4.

Element x :	1	i	-1	$-i$	j	k	$-j$	$-k$
Image $T(x)$:	1	j	-1	$-j$	k	i	$-k$	$-i$

Now confirm that the computations we did in the **Before the Lab** section are accurate by entering the information for S_3 in *ESG*.

4. We have constructed one automorphism of Q_4 and one nontrivial endomorphism of Q_4 that is not an automorphism. There are 24 automorphisms of Q_4. Explain why. Then find two other nontrivial endomorphisms of Q_4 which are not automorphisms. Write down the kernel and image of each endomorphism.

5. Find three nontrivial endomorphisms of S_3 which are not automorphisms. Write down the kernel and image of each. How many automorphisms of S_3 are there? Explain.

6. There are eight automorphisms of D_4. Explain why, and find two of them. What is the kernel of each? There are 27 nontrivial endomorphisms of D_4. Find one whose image is isomorphic to Z_2 and one whose image is larger than Z_2. Write down the image and kernel of each.

7. Find the five nontrivial endomorphisms of D_5 which are not automorphisms. Then find the seven nontrivial endomorphisms of D_7 which are not automorphisms. Write down the kernel and image of each.

8. Make a conjecture about the number of automorphisms of D_n or the number of endomorphisms of D_n.

9. True or false? Every subgroup of a group G is the image of some endomorphism of G. Give evidence to support your answer. If you answered false, try to write an alternate statement which might be true.

10. True or false? Every subgroup of a group G is the kernel of some endomorphism of G. Give evidence to support your answer. If you answered false, try to write an alternate statement which might be true.

11. If T is any endomorphism of a group G, prove that $\ker(T)$ is a subgroup of G.

12. If T is any endomorphism of a group G, prove that $\operatorname{im}(T)$ is a subgroup of G.

Chapter 11

The Inner Automorphisms of a Group

In Lab 10, you were introduced to automorphisms of a group G. In this lab, you will learn about the interesting relationship between the group G and a special collection of automorphisms called **inner automorphisms.**

11.1 Before the Lab

Definition 11.1: Let G be a group and $y \in G$. Define $T_y : G \to G$ by $T_y(x) = yxy^{-1}$ for all $x \in G$. We say that T_y is an **inner automorphism** of G. The collection of all inner automorphisms of G is denoted by $\text{Inn}(G)$.

For homework, you will prove that T_y is an automorphism of G.

We will start by computing a few of the inner automorphisms of S_3 by hand. For example, for the function T_{m_1}, we must calculate $T_{m_1}(x) = m_1 x m_1^{-1} = m_1 x m_1$ for every $x \in S_3$. Thus,

$$
\begin{aligned}
T_{m_1}(1_T) &= m_1 1_T m_1 = 1 & T_{m_1}(m_1) &= m_1 m_1 m_1 = m_1 \\
T_{m_1}(r_1) &= m_1 r_1 m_1 = r_2 & T_{m_1}(m_2) &= m_1 m_2 m_1 = m_3 \\
T_{m_1}(r_2) &= m_1 r_2 m_1 = r_1 & T_{m_1}(m_3) &= m_1 m_3 m_1 = m_2.
\end{aligned}
$$

Confirm by inspection that this rule is one-to-one and onto.

1. Do the necessary calculations by hand to determine the rules for T_{r_2} and T_{m_3}. Are these one-to-one and onto?

We will use *ESG* to determine the rest of the inner automorphisms of S_3.

11.2   In the Lab

We can generate inner automorphisms of a group G by using the conjugation command. Choose option 4 (Conjugates) from the **Group Properties Menu** and then action 2 (conjugation by a specific element). By asking for a value of y, *ESG* is asking to you to specify the y value for T_y.

2. Check the work you did for T_{r_2} and T_{m_3} using the computer. Then generate the other three inner automorphisms of S_3. Record the information for the six automorphisms in a table. Are they all distinct?

3. Use *ESG* to generate the inner automorphisms for D_n, $4 \leq n \leq 8$. Record the rule of each function in a chart.

4. Does each element of D_n generate a distinct inner automorphism? Give evidence for your answer.

5. There are $2n$ elements in D_n. Make a conjecture about the size of $\text{Inn}(D_n)$.

6. Calculate $\text{Inn}(G)$, where $G = Q_4$ and A_4. Record the rule of each function.

7. If G is an arbitrary group, can $T_x = T_y$ when $x \neq y$? Justify your answer using your data.

8. With *ESG*, the color blocks show that an inner automorphism of a group G always maps cosets of the center of G to cosets of the center. Explain in your own words why this must be true.

9. Based on your data, make a conjecture about the size of $\text{Inn}(G)$ when G is a finite group.

You will prove the following theorem for homework:

Theorem 11.1: Let G be a group. Then $\text{Inn}(G)$ and $\text{Aut}(G)$ are groups.

10. Make a group table for the elements of $\text{Inn}(S_3)$. The computer won't help you here—each element of $\text{Inn}(S_3)$ is a function. What is the group operation? The group $\text{Inn}(S_3)$ has six elements; we know there are two nonisomorphic groups of order 6. Can you identify to which one $\text{Inn}(S_3)$ must be isomorphic?

11. Try to identify $\text{Inn}(D_n)$, $4 \le n \le 8$. You can probably make a conjecture about the group without constructing the whole Cayley table. Then make a conjecture about the structure of $\text{Inn}(D_n)$ for any n.

12. Compute $\text{Inn}(Z_n)$ for several values of n. Then make a conjecture about the structure of $\text{Inn}(Z_n)$ for any n.

13. Describe the elements of $\text{Aut}(Z_5)$ and $\text{Aut}(Z_8)$. Then make a conjecture about the structure of $\text{Aut}(Z_n)$ for any n.

11.3 Further Work

14. Let G be a group and $y \in G$. Prove that T_y is an automorphism of G. (Hint: be sure to show first that T_y satisfies the definition of a homomorphism.)

15. Prove Theorem 11.1.

16. Characterize the groups G for which $\text{Inn}(G) = \{1_G\}$.

17. By problem 14 and theorem 11.1, we see that $\text{Inn}(G)$ is a subgroup of $\text{Aut}(G)$ for any group G. Is it possible for $\text{Inn}(G) = \text{Aut}(G)$? Justify your answer.

18. Prove that $\mathrm{Inn}(G)$ is a *normal* subgroup of $\mathrm{Aut}(G)$.

19. Make a conjecture about the relationship between G and $\mathrm{Inn}(G)$ for an arbitrary group G.

Chapter 12

Is a Normal Subgroup Always the Kernel of an Endomorphism?

You know that the kernel of an endomorphism of a group G is always a normal subgroup of G. But is the converse of this theorem true? In this lab, you'll generate examples to investigate this question and make a conjecture based on the evidence.

12.1   Before the Lab

Recall these definitions made in Lab 10:

Definition 10.1: Let G and K be groups. A function $T : G \to K$ is called a group **homomorphism** if $T(xy) = T(x)T(y)$ for every $x, y \in G$. When $K = G$, we say that T is a group **endomorphism**.

Definition 10.2: The **kernel** of a homomorphism T is the set

$$\ker(T) = \{x \in G \mid T(x) = 1_G\}.$$

Definition 10.3: The **image** of a homomorphism T is the set

$$\operatorname{im}(T) = \{y \in K \mid \exists x \in G \text{ with } T(x) = y\}.$$

As in Lab 10, the structure of ESG will require us to consider group endomorphisms instead of general homomorphisms. You proved for homework in Lab 10 that when T is an endomorphism, $\ker(T)$ and $\operatorname{im}(T)$ are subgroups of the group G.

Your textbook proves the following:

Theorem 12.1: Let G and K be groups. If $T : G \to K$ is a homomorphism, then $\ker(T)$ is a normal subgroup of G.

Suppose that N is a normal subgroup of G. Could we find an *endomorphism* $T : G \to G$ with $\ker(T) = N$?

This goal of this lab is to see if the answer to this question could be "yes." If you are able to construct such an endomorphism for every group you test, then you might be led to restate this question as a firm conjecture. If your work leads to a counterexample, then you need to consider how to create a reasonable conjecture.

The computer will use the following two rules to compute the endomorphism once you choose some initial values for the function:

    i. **Power rule:** If you set $T(a) = b$, then the computer calculates that $T(a^r) = b^r$.

    ii. **Homomorphism rule:** If you set $T(a_1) = b_1$ and $T(a_2) = b_2$, then the computer calculates that $T(a_1 a_2) = b_1 b_2$.

12.2   In the Lab

The example below will illustrate how you may want to approach the exercises in the lab. It is similar to an example worked out in Lab 10; some of the details of that example have been repeated for clarity. Enter the information into your computer as you read through the example.

Let $G = Q_4$, the quaternions. Rename and reorder the elements as below in order to obtain the familiar table:

1	A	B	C	D	E	F	G
1	i	-1	$-i$	j	k	$-j$	$-k$

Let $N = <i> = \{1, i, -1, -i\}$. Since N is cyclic, setting $T(i) = 1$ will guarantee that $N \subset \ker(T)$. The computer will calculate

$$T(-1) = T(i^2) = 1^2 = 1$$
$$T(-i) = T(i^3) = 1^3 = 1$$
$$T(1) = T(i^4) = 1^4 = 1.$$

The element $j \notin N$, so we would like that $T(j) \neq 1$ (or else the kernel of T will not be precisely N). But $j^2 = -1$, and we have already calculated $T(-1)$, so we need to be careful about our choice of $T(j)$. Let $T(j) = -1$. The computer will calculate

$$T(-1) = T(j^2) = (-1)^2 = 1$$
$$T(-j) = T(j^3) = (-1)^3 = -1$$
$$T(1) = T(j^4) = (-1)^4 = 1.$$

This agrees with our previous work. We actually don't need to make any more choices; the computer will use the homomorphism rule to calculate the rest of the values. We know that $k = ij$, so

$$T(k) = T(ij) = T(i)T(j) = (1)(-1) = -1.$$

Once again, the power rule gives

$$T(-1) = T(k^2) = (-1)^2 = 1$$
$$T(-k) = T(k^3) = (-1)^3 = -1$$
$$T(1) = T(k^4) = (-1)^4 = 1.$$

This agrees with our previous work and we have a genuine endomorphism T with $\ker(T) = N$. Confirm that the computer gives the same values that we have calculated above, and that $\ker(T) = N$.

1. Answer the questions below for the following groups: Z_5, D_4, Q_4, D_5, Z_{12}, D_6, A_4 and Q_6. You will need to have available an annotated subgroup lattice for each group.

    (a) For every normal subgroup N of G, try to construct an endomorphism $T : G \to G$ with $\ker(T) = N$.

(b) If you can find such an endomorphism, enter your information in a chart which has the headings:

N rule of the endomorphism $\text{im}(T)$ G/N

(c) If you cannot find an endomorphism, try to discover why not. In other words, what is the obstruction to creating such an endomorphism?

12.3   Further Work

2. What is your answer to the question we asked at the beginning of the lab: Given N a normal subgroup of G, does there exist an endomorphism $T : G \to G$ with $\ker(T) = N$?

3. If you answered "no" to problem 2, write a conjecture that you believe is true.

4. In several sentences, explain the information in the charts and justify your conjecture.

Chapter 13

The Class Equation

Simple counting arguments can lead to important results in mathematics. The **Class Equation** is one informative way to count the elements of a group.

13.1   Before the Lab

Definition 13.1: The **centralizer** of an element $g \in G$ is the collection of elements in G which commute with the element g. That is,

$$C_G(g) = \{x \in G \mid xg = gx\}.$$

For homework, you will prove that the centralizer $C_G(g)$ is a subgroup of G.

Definition 13.2: Suppose that G is an arbitrary group and let $g \in G$. We can conjugate g by every element x of G, xgx^{-1}. The collection of elements of G that we obtain is called the **conjugacy class** of g, denoted by class(g).

Theorem 13.1: Let G be an arbitrary group, and suppose that $g, h \in G$. Define the relation $\sim$ as follows: $g \sim h$ if and only if there exists an element $x \in G$ such that $xgx^{-1} = h$. Then $\sim$ is an equivalence relation on the set of elements of G.

We will be using *ESG* to compute both the centralizer of elements and the conjugacy classes in a group. First, however, we will work out the example of S_3 by hand. Let's start with the element $r_1 \in S_3$. We need to calculate xr_1x^{-1} for every $x \in S_3$.

$$1_T r_1 1_T = r_1 \qquad m_1 r_1 m_1 = r_2$$
$$r_1 r_1 r_2 = r_1 \qquad m_2 r_1 m_2 = r_2$$
$$r_2 r_1 r_1 = r_1 \qquad m_3 r_1 m_3 = r_2$$

These computations show that the elements r_1 and r_2 are conjugate to r_1, so class$(r_1) = \{r_1, r_2\}$ is a cell of a partition of the elements of S_3.

1. For the remaining elements $g \in S_3$, compute xgx^{-1} as x runs through all of the elements of S_3. What are the distinct conjugacy classes of S_3?

To determine the centralizer $C_{S_3}(r_1)$, we need to check the products xr_1 and r_1x for each $x \in S_3$. We have $1_T r_1 = r_1 1_T, r_1 r_1 = r_1 r_1$, and $r_2 r_1 = r_1 r_2$, but $m_1 r_1 \neq r_1 m_1, m_2 r_1 \neq r_1 m_2$, and $m_3 r_1 \neq r_1 m_3$. Therefore $C_{S_3}(r_1) = \{1, r_1, r_2\}$.

2. Calculate $C_{S_3}(g)$ for the remaining elements $g \in S_3$.

3. Explain what happens when you compute the conjugacy class of an element in an abelian group. What is the centralizer of an element in an abelian group?

4. Prepare a chart with the following headings to record data in the lab:

   group G element g class(g) #(class(g)) $C_G(g)$ #($C_G(g)$)

13.2   In the Lab

5. Complete the chart for the groups: S_3, D_4, Q_4, D_5, D_6, A_4, Q_6, D_7, M, D_8 and Q_8. Use option 4 (Conjugates) in the **Group Properties Menu**. *ESG* will compute both the conjugacy class of an element A, denoted by class(A), and the centralizer of the element A.

6. Is the conjugacy class $C_G(g)$ of an element $g \in G$ a subgroup of G? Give evidence for your answer.

7. Is the centralizer $C_G(g)$ of an element $g \in G$ a normal subgroup of G? Justify your answer.

8. Is it possible for class$(g) = \{g\}$, for some $g \in G$, $g \neq 1_G$? Explain your answer.

9. Suppose that two elements $x, y \in G$ are contained in the same conjugacy class. That is, class(x) = class(y). Is it true that $C_G(x) = C_G(y)$? Give evidence to support your answer.

Since conjugation is an equivalence relation on the group G, we can count the elements of G by counting the elements in the distinct conjugacy classes.

10. Using the data in your chart, give a formula which computes $\#(\text{class}(g))$ for some $g \in G$.

11. If $g \in Z(G)$ (the center of G), what is $\#(\text{class}(g))$?

13.3 Further Work

12. Combining your answers from questions 10 and 11, write a formula which counts the number of elements in an arbitrary group G.

13. Prove that the centralizer $C_G(g)$ of an element $g \in G$ is a subgroup of G.

14. Prove Theorem 13.1.

Chapter 14

Conjugate Subgroups

Inner automorphisms, studied in Lab 11, can be used to create equivalence classes of subgroups of a group. This lab will give you additional insight into the relationships among subgroups of a group and will provide background material for Lab 15, on the Sylow Theorems.

14.1   Before the Lab

Recall the following definition made in Lab 11:

Definition 11.1: Let G be a group and H and K be subgroups of G. An **inner automorphism** $T_y : G \rightarrow G$ is defined by $T_y(x) = yxy^{-1}$, where $g \in G$.

Definition 14.1: We say that H is **conjugate** to K if there exists an inner automorphism T_y for some $y \in G$ such that $T_y(H) = K$.

As in the previous labs, we will work with the example of S_3 by hand. Consider the subgroups

$$H_1 = \{1_T, m_1\} \quad H_2 = \{1_T, m_2\} \quad H_3 = \{1_T, m_3\} \quad H_4 = \{1_T, r_1, r_2\}.$$

To discover which subgroups are conjugate to H_1, we need to compute xH_1x^{-1} for every $x \in S_3$.

$$1_T H_1 1_T = \{1_T, 1_T m_1 1_T\} = \{1_T, m_1\} = H_1$$
$$r_1 H_1 r_2 = \{r_1 1_T r_2, r_1 m_1 r_2\} = \{1_T, m_2\} = H_2$$
$$r_2 H_1 r_1 = \{r_2 1_T r_1, r_2 m_1 r_1\} = \{1_T, m_3\} = H_3$$
$$m_1 H_1 m_1 = \{m_1 1_T m_1, m_1 m_1 m_1\} = \{1_T, m_1\} = H_1$$
$$m_2 H_1 m_2 = \{m_2 1_T m_2, m_2 m_1 m_2\} = \{1_T, m_3\} = H_3$$
$$m_3 H_1 m_3 = \{m_3 1_T m_3, m_3 m_1 m_3\} = \{1_T, m_2\} = H_2$$

That is, H_1 is conjugate to H_2 and H_3.

1. Do the same computations for the subgroups H_2, H_3 and H_4. Record this information on your subgroup lattice of S_3. (For example, underline all conjugate subgroups in the same color.)

For homework, you will prove the following:

Theorem 14.1: If N is a normal subgroup of a group G, then $T_y(N) = N$ for every $y \in G$.

Which subgroup of S_3 satisfies the theorem?

Theorem 14.2: Let $\mathfrak{F}$ denote the collection of subgroups of a group G. Conjugation is an equivalence relation on $\mathfrak{F}$.

2. Write down the cells of this equivalence relation on the set of subgroups (including the improper and trivial subgroups) of S_3.

Be sure to bring your subgroup lattices for the groups D_4, Q_4, D_5, D_6, A_4, Q_6, D_7, M, D_8, and Q_8 to the lab.

14.2   In the Lab

Because of Theorem 14.1, *ESG* will calculate the conjugates only of subgroups that are not normal.

3. For each of the groups $D_4, Q_4, D_5, D_6, A_4, Q_6, D_7, M, D_8$, and Q_8, determine which subgroups are conjugate. Then record this information on your subgroup lattices.

4. Write down one conjecture about conjugate subgroups which seems reasonable based on the data generated from this lab.

5. Write one conjecture about the conjugate subgroups of D_n.

14.3 Further Work

6. Prove Theorem 14.1.

7. Prove Theorem 14.2.

Chapter 15

The Sylow Theorems

What can you say about the structure of a group if you just know the order of the group? "Not much," you say, if the order is not a prime number. The powerful Sylow Theorems will change your response. They give you information about certain subgroups of a group G, which in some cases will completely determine the structure of G. You will see this structure most clearly when $|G| = p^r n$, with $(p, n) = 1$. Because of the limitations of ESG, you will be working with the 5 nonisomorphic groups of order 12.

15.1 Before the Lab

Definition 15.1: We say that K is a p–**group** if every element of K has order that is a power of p. If $K \subset G$ and K is a p–group, we say that K is a p–**subgroup** of G.

Consider the subgroup lattice of A_4:

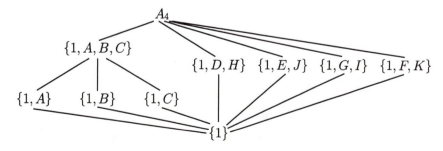

The 2–subgroups of A_4 are $\{1, A\}$, $\{1, B\}$, $\{1, C\}$, $\{1, A, B, C\}$ and the 3–subgroups of A_4 are $\{1, D, H\}, \{1, E, J\}, \{1, G, I\}$, and $\{1, F, K\}$. We can arrange these into ascending chains, for example

$$\{1\} \subset \{1, A\} \subset \{1, A, B, C\} \subset A_4 \qquad (15.1)$$

$$\{1\} \subset \{1, D, H\} \subset A_4 \qquad (15.2)$$

In (15.1), we notice that $\{1, A, B, C\}$ is the largest proper 2–subgroup. That is, A_4 does not have a 2–subgroup which we can place between $\{1, A, B, C\}$ and A_4 in the chain. Similarly, $\{1, D, H\}$ is the largest proper 3–subgroup of A_4. We give these maximal subgroups a name:

Definition 15.2: Let G be a group with order $p^r n$, where $(p, n) = 1$. We say that a maximal proper p–subgroup is a **Sylow p–subgroup**. Let n_p denote the number of Sylow p–subgroups of G.

1. Write all other possible ascending chains of 2–subgroups and 2–subgroups of A_4. Then identify all Sylow 2–subgroups and Sylow 3–subgroups of A_4.

This lab uses subgroup lattices you have constructed previously. It may be possible for you to complete the lab without any additional work on the computer. Or, you may find that you need the computer to construct a lattice or fill in details on a lattice you have previously constructed.

15.2   In the Lab

Be sure to bring to the lab the subgroup lattices of Z_{12}, $Z_6 \times Z_2$, D_6, A_4, Q_6 in addition to those of Z_6, S_3, Z_{10} and D_5.

2. For the other groups of order 12, identify all possible 2– and 3–subgroups. Then write down all possible ascending chains of 2–subgroups and 3–subgroups. Identify all the Sylow 2–subgroups and Sylow 3–subgroups for each group.

3. For each group of order 12, can you say there is a unique Sylow 2–subgroup? A unique Sylow 3–subgroup?

4. For each group of order 12, what is the order of a Sylow 2–subgroup? A Sylow 3–subgroup?

5. For a group with order $p^r n$, with $(p, n) = 1$, write a conjecture about the order of a Sylow p–subgroup.

6. Calculate n_p, $p = 2, 3$, for the 5 groups of order 12. Then complete the following sentences.

Suppose that $\mid G \mid = 12$.

   (a) Then G has either ___ or ___ Sylow 3–subgroups, which are all of order ___.

   (b) Then G has either ___ or ___ Sylow 2–subgroups, which are all of order ___.

7. You should not have any blanks filled with the number 0 in problem 6. Do you think that, for some group of order other than 12, we could have $n_p = 0$? Justify your answer.

8. Let G be a group with order $p^r n$, where $(p, n) = 1$. Write a conjecture that relates n_p, the number of Sylow p–subgroups, to $\mid G \mid$.

9. Write a conjecture about the Sylow p–subgroups of a group, if $n_p > 1$.

10. Write a conjecture about the unique Sylow p–subgroup of a group when $n_p = 1$.

11. Now consider the groups of order 6 , Z_6 and S_3, and order 10, Z_{10} and D_5. Do these groups satisfy your conjectures? If not, refine the conjectures made in questions 5, 8, 9, and 10.

15.3   Further Work

12. There are 230 nonisomorphic groups of order 96, but only 1 group of order 95. Note that neither 230 nor 95 is a prime number. Without writing a careful proof, think about why the first sentence could be true. Explain your reasoning in several complete sentences.

Appendix A

Table Generation Menu of *ESG*

In order to use *ESG* you must choose an operation table upon which to base all calculations. If C appears in such a table at the intersection of the A-row and the B-column we call C the "product" or "composite" of A and B and write $C = A * B$. There are four different ways to choose a table.

Please select an option (1-5) from the following:

1. Operation Table from the Sample Library
2. Random Operation Table
3. User-Defined Operation Table
4. Operation Table from the Group Library
5. Quit this Program

Appendix B

Sample Library of *ESG*

This is option 1 from the **Table Generation Menu**. Many of the sample tables give multiplication of integers modulo n. Z_n has all remainders 0 through $n - 1$, while U_n has only the invertible elements of Z_n. The size of each sample is listed first.

1. $9 - Z_9$
2. $11 - Z_{11}$
3. $12 - Z_{12}$
4. $13 - Z_{13}$
5. $14 - Z_{14}$
6. $15 - Z_{15}$
7. $16 - Z_{16}$
8. $6 - U_{18}$
9. $8 - U_{20}$
10. $8 - U_{24}$
11. $8 - U_{30}$
12. $10 - U_{22}$
13. $12 - U_{21}$
14. $12 - U_{26}$
15. $12 - U_{28}$
16. $12 - U_{36}$
17. $12 - U_{42}$
18. $16 - U_{17}$
19. $16 - U_{32}$
20. $16 - U_{34}$
21. $16 - U_{40}$
22. $16 - U_{48}$
23. $16 - U_{60}$
24. $9 -$ Squares in U_{19}
25. $9 -$ Squares in U_{63}
26. $14 -$ Squares in U_{29}
27. $15 -$ Squares in U_{77}
28. $16 -$ Squares in U_{65}
29. $16 - XOR$ in a 4-set
30. $16 - 2 \times 2$ Matrices Mod 2
31. $16 -$ Cayley's Octonions
32. $3 -$ Quasigroup
33. $5 -$ Loop
34. $6 -$ Commutative Loop
35. $8 -$ Maps of ABC into AB
36. $12 -$ Divisors of 60

Appendix C

Group Library of *ESG*

This is option 4 from the **Table Generation Menu.** The four-digit codes below refer to group tables. The first two digits of the code give the size of the group. Common names are given with most groups.

$0301 - Z_3$ Cyclic

$0401 - Z_4$ Cyclic

$0402 - Z_2 \times Z_2$, Four-Group V

$0501 - Z_5$ Cyclic

$0601 - Z_6$ Cyclic

$0602 - D_3$ Dihedral / S_3 Symmetric

$0701 - Z_7$ Cyclic

$0801 - Z_8$ Cyclic

$0802 - Z_4 \times Z_2$

$0803 - Z_2 \times Z_2 \times Z_2$ Elementary

$0804 - D_4$ Dihedral / Octic

$0805 - Q_4$ Dicyclic / Quaternion

$0901 - Z_9$ Cyclic

$0902 - Z_3 \times Z_3$ Elementary

$1001 - Z_{10}$ Cyclic

$1002 - D_5$ Dihedral

$1101 - Z_{11}$ Cyclic

$1201 - Z_{12}$ Cyclic

$1202 - Z_6 \times Z_2$

$1203 - D_6$ Dihedral

$1204 - A_4$ Alternating Subgroup of S_4

$1205 - Q_6$ Dicyclic

$1301 - Z_{13}$ Cyclic

$1401 - Z_{14}$ Cyclic

$1402 - D_7$ Dihedral

$1501 - Z_{15}$ Cyclic

$1601 - Z_{16}$ Cyclic

$1602 - Z_8 \times Z_2$

$1603 - Z_4 \times Z_4$

$1604 - Z_4 \times Z_2 \times Z_2$

$1605 - Z_2 \times Z_2 \times Z_2 \times Z_2$ Elementary

$1606 - D_4 \times Z_2$

$1607 - Q_4 \times Z_2$

$1608 -$ A Subgroup of $GL_2(Z_5)$

$1609 -$ Sylow 2-Sg of $SL_2(Z_4)$

$1610 -$ Semidirect product of Z_4 by Z_4

$1611 -$ A Subgroup of $GL_2(Z_5)$, M

$1612 - D_8$ Dihedral

$1613 -$ Sylow 2-Sg of $GL_2(Z_3)$

$1614 - Q_8$ Dicyclic

Appendix D

Group Properties Menu

Please select an option (1-7) from the list below.

1. Powers and Order
2. Commuting Elements
3. Subgroups and Cosets / Quotients
4. Conjugates

5. Endomorphisms
6. Go to Property Checks Menu
7. Table Alterations